Sticky

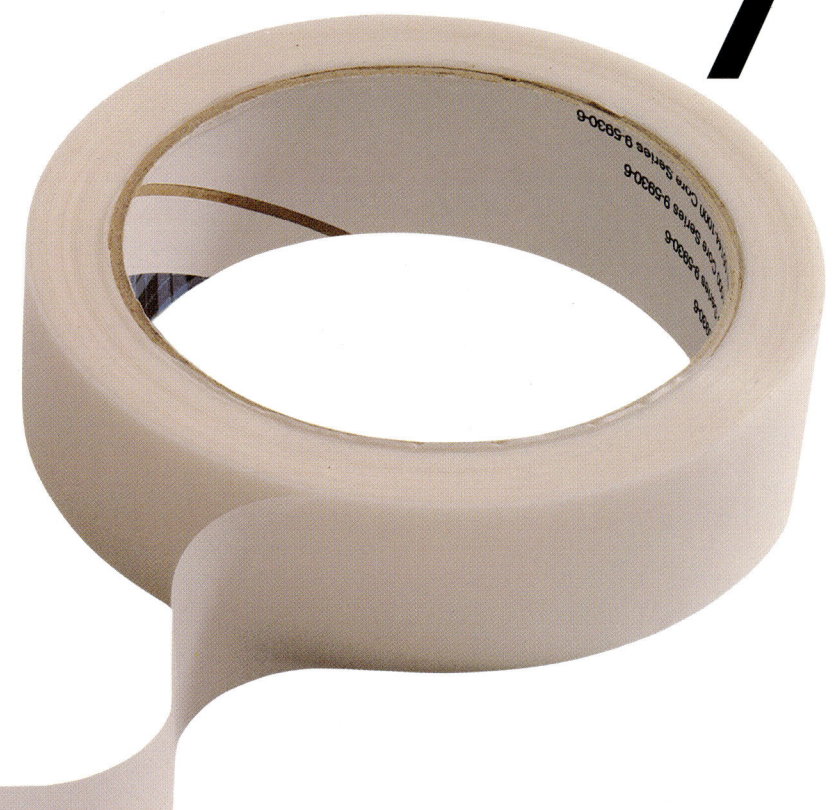

RIGBY
INTERACTIVE
LIBRARY

© 1996 Rigby Education
Published by Rigby Interactive Library,
an imprint of Rigby Education,
division of Reed Elsevier, Inc.
500 Coventry Lane,
Crystal Lake, IL 60014

All rights reserved. No part of this publication may be reproduced or transmitted in any form or by any means, electronic or mechanical, including photocopying, recording, taping, or any information storage and retrieval system, without permission in writing from the publisher.

Cover designed by Herman Adler Design Group
Designed by Heinemann Publishers (Oxford) Ltd
Printed in China

00 99 98 97 96
10 9 8 7 6 5 4 3 2 1

Library of Congress Cataloging-in-Publication Data

Warbrick, Sarah, 1964-
What is sticky?
   p. cm. -- (What is--?)
Summary: Presents the physical property of adhesion in everyday objects.
ISBN 1-57572-052-3 (library)
1. Matter--Properties--Juvenile literature. 2. Adhesives--Juvenile literature. [1.Matter--Properties. 2. Adhesives.]
 I. Title. II. Series: Warbrick, Sarah, 1964- What is--?
QC173.36.W444 1996
620.1'1292--dc20   95-41111
                                                                             CIP
                                                                              AC

Acknowledgments
The publishers would like to thank Toys Я Us Ltd,
the world's biggest toy megastore ,
for the kind loan of equipment and materials
used in this book.

Special thanks to Jodie, Katie, Michael, Nadia, Rose,
and Winnie who appear in the photographs

Photographs: Bruce Coleman pp10, 11; NHPA pp8, 9;
other photographs by Trevor Clifford
Commissioned photography arranged by Hilary Fletcher

There are sticky things all around us.
Sticky things can be fun.
Sticky things can be useful.
Sticky things can also be messy!

This book shows you what is sticky.

These things look different.
What differences can you see?

In one way they are all the same.
They are all sticky.

Glue is very sticky.

Sarah can stick things together with it.

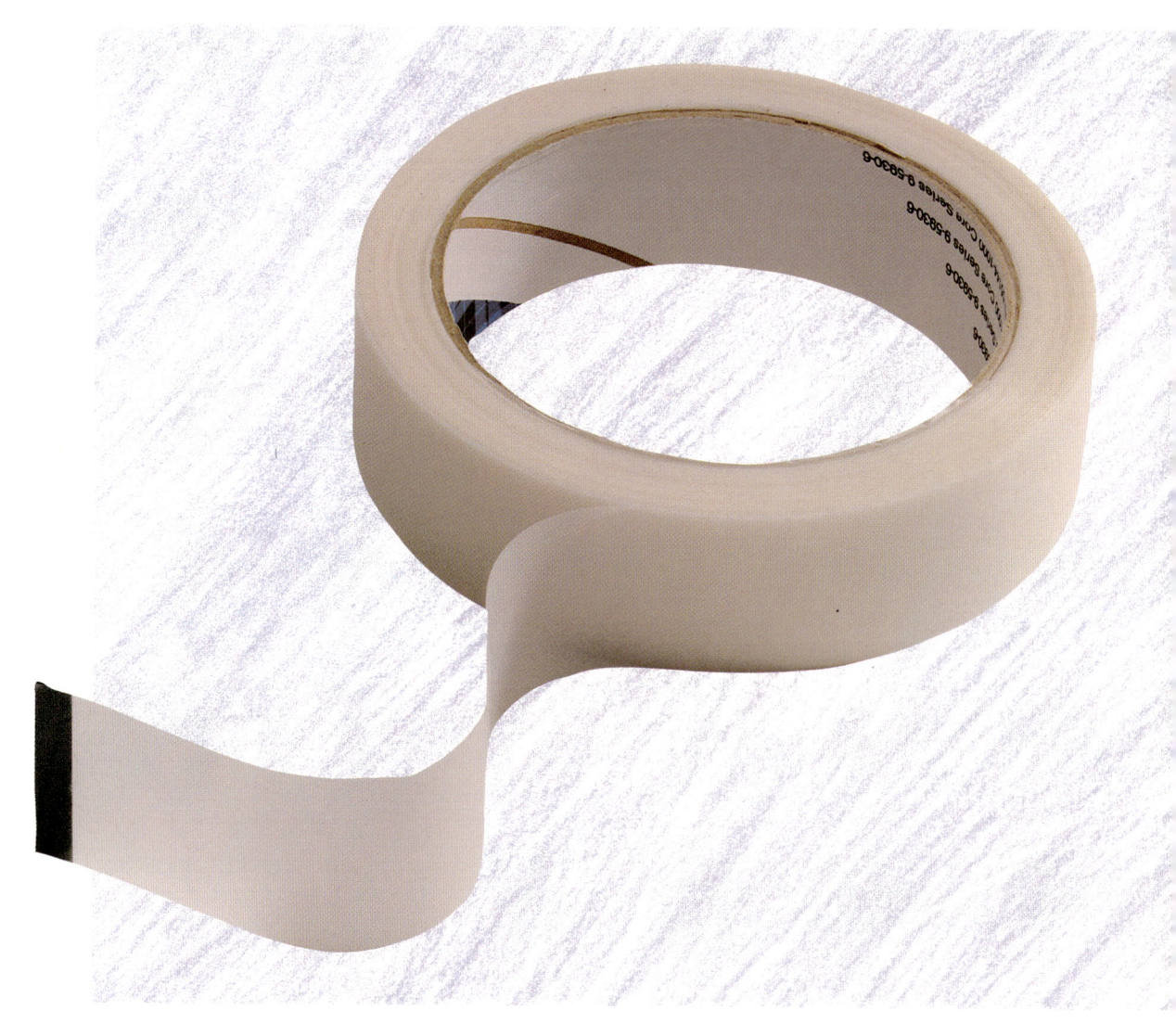

Sticky tape has glue on it.

What do you use tape for?

Some seeds have tiny hooks on their surfaces.

They can stick to your clothes and to animals' fur.

The pollen from this flower sticks to the bee's legs.

The bee carries the pollen to another flower and it becomes unstuck.

These shoes have a special sticky fastener.

This makes them easy to put on and take off.

Mud can stick to your shoes.

But it can easily become unstuck on the carpet!

Pizza dough is sticky.

But look at it now that it's cooked.

Sticky toffee can stick to your teeth.

Be careful!
Your teeth might come unstuck!

What is sticky here?